Puzzles 101

A Puzzlemaster's Challenge

Nobuyuki Yoshigahara

Translated by
Richard Weyhrauch
and Yasuko Weyhrauch

A K Peters
Natick, Massachusetts

Editorial, Sales, and Customer Service Office

A K Peters, Ltd.
63 South Avenue
Natick, MA 01760
www.akpeters.com

Library of Congress Cataloging-in-Publication Data

Yoshigahara, Nobuyuki, 1936-
    [Chocho nanmon suri pazuru. English]
    Puzzles 101: a puzzlemaster's challenge / Nobuyuki Yoshigahara ; translated by Richard Weyhbrach and Yasuko Weyhbrach.
      p. cm.
    ISBN 1-56881-177-2
    1. Mathematical recreations. I. Title: Puzzles one hundred and one. II. Title: Puzzles one hundred one. III. Title.

QA95.Y6713 2003
793.74–dc22

2003062426

Printed in Canada
08 07 06 05 04   10 9 8 7 6 5 4 3 2 1

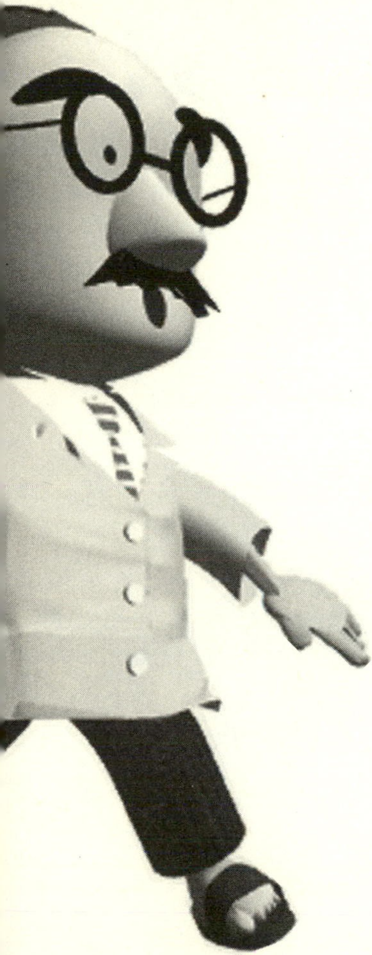

Puzzles 101

Table of Contents

Preface   ix

1. Two Different Lengths   1, 67
2. Crossing the River   2, 67
3. Making a Die   3, 68
4. What Comes Next?   3, 68
5. Diamond Puzzle   4, 69
6. The Water Puzzle   5, 69
7. The Opposite of Puzzle 6   6, 70
8. Three Times More   6, 70
9. Triangular Arithmetic   7, 71
10. Triangular Solitaire   8, 71
11. The Same Area   9, 72
12. Lose Those Squares   10, 73
13. Two Choice Question   10, 73
14. Triangulation   11, 74
15. Rock, Paper, Scissors   11, 74
16. Four Solutions of Four Pieces I   12, 75
17. Four Solutions of Four Pieces II   12, 75
18. Square   13, 76
19. Simultaneous Equations   13, 76
20. Even Number Sequence   14, 77
21. Age Guessing   15, 77
22. Divide by Four   15, 78
23. Can You?   16, 78
24. To Have or Not to Have   16, 79
25. Pentominos   17, 79
26. Pandigital (Komachi) Fraction I   18, 80
27. Pandigital (Komachi) Fraction II   18, 80
28. Triangulate a Square   19, 81
29. Challenge by Don Knuth   20, 81
30. Square Numbers   21, 82
31. Interlocking Pieces?   21, 83
32. Change of Heart   22, 83
33. Fill It Up With Coins   22, 84

34. LYM   23, 84
35. Bug House   23, 85
36. What's Missing?   24, 85
37. Connecting the Dots   25, 86
38. Equal Distance   25, 86
39. Palindrome Time   26, 87
40. Pandigital (Komachi) Time   27, 87
41. TROMINO   27, 88
42. Switching Trains   28, 88
43. Building Highways   28, 89
44. Congruent Parts   29, 89
45. All Gone?   30, 90
46. Change Direction   30, 90
47. Arrest Him   31, 91
48. Pandigital Olympic Circles   31, 91
49. Massive Multiplication   32, 92
50. Cut and Rearrange a Cross   32, 92
51. The Curious Door Sign   33, 93
52. Triangle Struggle   33, 93
53. Place the Coins   34, 94
54. Cut and Paste I   34, 94
55. Cut and Paste II   35, 95
56. Trick Name Card I   35, 95
57. Trick Name Card II   36, 95
58. Uniform Division   36, 96
59. Three Contacting Matchsticks   37, 97
60. Find the Rule   37, 98
61. Typewriter   38, 98
62. Seven Straight Lines   38, 98
63. Four and a Half Tatami Mats   39, 99
64. Five Consecutive Numbers   40, 100
65. Chocolate   40, 100
66. A Go Stone Puzzle   41, 100
67. Coins in Two Dimensions   41, 101

68. Make a Square 42, 107
69. What Comes Next? II 42, 102
70. Equally Divided 43, 102
71. Wild Pig Maze 43, 103
72. FlipIt 44, 104
73. A Shining Star 45, 105
74. Escape from the Tower 45, 105
75. A Broken Calculator I 46, 106
76. A Broken Calculator II 46, 106
77. Switching Frogs 47, 107
78. All Different 48, 107
79. Odd Numbers and Even Numbers 49, 108
80. Magic Trick 50, 108
81. Time Calculation Using Pandigitals (Komachi) 51, 109
82. Time Calculation Using Pandigitals Plus 0 (Komachi) 51, 109
83. How Many Loops? 52, 110
84. Up and Down Maze 53, 111
85. Jumping Go Stones 54, 112
86. Moving Matchsticks 55, 112
87. One Place the Same 56, 113
88. The Same Product 57, 113
89. No Squares 57, 114
90. Taking a Walk 58, 114
91. From 1 to N 60, 116
92. Self-Satisfaction 59, 115
93. From 1 to 10 60, 116
94. Height and Weight 60, 116
95. Tail Becomes Head 61, 117
96. Is the Opposite Also True? 61, 117
97. Multi-Level Crossing 62, 118
98. Squeeze-In! I 63, 119
99. Squeeze-In! II 63, 120
100. UN-Balance 64, 120
101. Surprised Solomon 65, 121

Preface

The famous German poet Hans Magnus Enzensberger concluded his essay on the isolation of mathematics in our society with an appeal for action: "The effort in question is nothing short of the achievement of mathematical literacy which will furnish our all to sluggish brains with a kind of athletic work-out and yield to us a variety of pleasure to which we are entirely unaccustomed." Those of us in the international puzzler's league have anticipated his advice and pursued our belief in the stimulating and entertaining quality of advanced puzzles. Puzzle creators and solvers are a close-knit community and our creations travel by word of mouth like the great poems, fairy tales, and sagas of the world. A gradual expansion into the wider community is changing the communication and the influence of our art. Judging by the growing popularity of books dealing with advanced puzzles, we are successful in spreading the word, and it has always made me proud and happy to see puzzles that I created appearing in such books. Suddenly, I began to worry: If I ever wanted to publish a book of my puzzles, would people not look at me like a thief? So I decided to start my own collection sooner rather than later, and I present, in this small volume, many puzzles that I invented and a number of puzzles created by my friends that I particularly like. I include those puzzles with their permission and mention their name as a small tribute. If you, dear Reader, find a puzzle without a name that you believe was made by someone else, please notify me or the publisher so that credit can be given where credit belongs.

My selection comes from a huge pile of puzzles and my choices were driven by my own taste and the desire to balance advanced and simple, but entertaining, puzzles to keep the reader interested and motivated. I don't believe that anybody could solve all these puzzles in one year; I certainly could not do it if I did not know the solutions already. But here is a condition that you must obey: Do not use a computer! Remember what Enzensberger said: "... an exercise to give our all too sluggish brains [not our computer] a ... work-out."

I used to be a maniac about solving hard puzzles and I had a lot of fun (and some frustration) doing it. Now, I have become a taskmaster who creates puzzles for others to enjoy and agonize over. Let me tell you that the reward of such agonizing is the great joy of "Eureka!" You may get addicted, but this kind of addiction will only help your brain, not destroy it.

Let me thank all who have allowed me to use their ideas, and let the joy of our readers be a reward for all of us who care about and puzzle over hard problems with elegant solutions.

PUZZLES

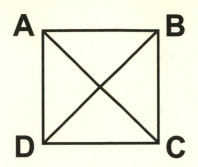

Let's look at a square of side-length 1. There are six straight lines that connect any two vertices of the square. The length of these lines is either

AB = BC = CD = DA = 1
or
AC = BD= √2.

Can you rearrange the four vertices so that they are on a plane and there are still only two different lengths connecting any two points? How many different arrangements can you think of? No two points are allowed to be in the same location.

(Puzzle by Dick Hess)

Crossing the River

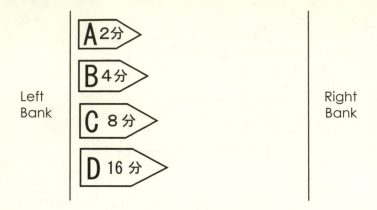

Left
Bank

Right
Bank

Four boats A–D are on the left bank of a river. To cross the river to the right bank, Boat A takes 2 minutes; Boat B takes 4 minutes; Boat C takes 8 minutes; and Boat D takes 16 minutes.

The four boats need to go to the right bank, but there is only one boatman. One boat can pull another boat, but it will take the same amount of time as the slower of the two boats to cross the river.

If the boatman uses one boat to pull another boat to the right bank and returns to the left bank in one boat, how many minutes will it take for him to move all the boats to the right bank? Find the shortest time (ignore the time he spends changing and connecting boats).

Making a Die

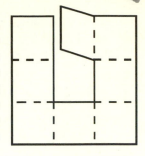

Fold a square into a 3 x 3 grid. Cut out the square in the middle and cut one section as shown here. You now have eight small squares. Can you fold this paper and make it into a cube? Since a cube only has six surfaces, you need to overlap some squares.

(Puzzle by Edward Hordern)

What Comes Next?

What comes after 66?

2, 4, 6, 30, 32, 34, 36, 40, 42, 44, 46, 50, 52, 54, 56, 60, 62, 64, 66, ?

A square is missing one-quarter of its area; we want to turn it into the diamond shape in Figure 1 by cutting it into two pieces and rearranging them.

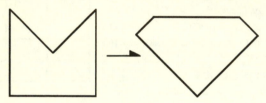

Figure 1.

In Figure 2, we cut the square into four pieces and rearranged them. Can you find a way to cut and rearrange fewer pieces and make a diamond?

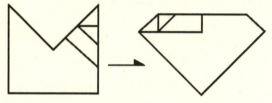

Figure 2.

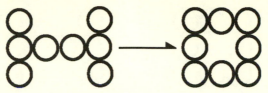

Eight coins are arranged to look like the letter H. By sliding one coin around at a time, we want to arrange the eight coins to look like the letter O. However, there is a strict rule: A moving coin needs to stop in a position where it touches at least two coins.

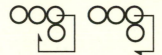

Figure 1.   Figure 2.

In Figure 1, the moving coin ends up touching two coins, but in the Figure 2, the moving coin ends up touching only one coin, so it is prohibited.

Figure 3.

In Figure 3, a moving coin is indeed touching two coins, but this coin will not stop right there—it will slide on; therefore, such a move is also prohibited.

Please get eight real coins and play with them. If you form the letter O in five moves, I will be really impressed.

By the way, I call this puzzle "The Water Puzzle" because it changes H to O (H_2O).

Puzzle 7 The Retaw Puzzle

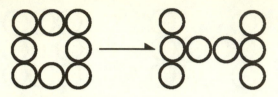

Continue from Puzzle 6. This time, each coin needs to be moved around so that the eight coins arranged as the letter O change into the letter H. The minimum number of moves is seven—this puzzle is much more difficult.

Puzzle 8 Three Times More

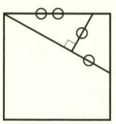

I divided a square into three parts. The lines labeled with one circle are the same length, and the length of the line with two circles is twice as long as the lines with one circle.

Get some paper, cut three squares, and then cut each of them into three pieces as shown here.

Now, using the nine pieces cut from the squares, create a square whose area is three times that of the original square.

This puzzle will be less difficult if you actually make the pieces.

Arrange the numbers 1 to n (where n is a "triangular" number) to form an equilateral triangle in such a way that the difference between two neighboring numbers is right below them. Figure 1 shows some samples:

Figure 1.

Now, arrange the numbers 1 to 15 to create such a triangle.

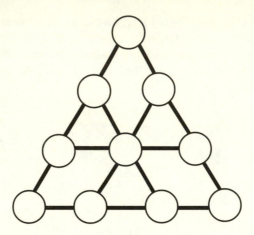

Connect ten circles in the form of an equilateral triangle as shown. Put one coin on each circle and follow these steps:

1. Remove one of the coins.
2. Pick another coin. If there is a coin on a circle adjacent to the coin you chose and the circle beyond that coin is empty, then you can jump over that coin and take it away.
3. Repeat Step 2. You are done when there is only one coin left.

Advanced: Try to solve this in as few steps as possible. Jumping continuously is counted as one step.

The Same Areas

Using three matchsticks, bisect the area of a triangle that is made up of three, four, and five matchsticks on each side, respectively. This can be solved as shown in Figure 1.

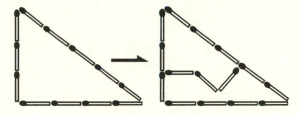

Now, decrease the number of matchsticks by one on each side and try to bisect the area of the triangle using only two matchsticks.

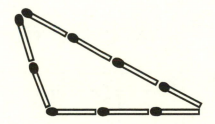

Puzzle 12 Lose Those Squares

Arrange 20 coins as in the picture. We can create 21 squares by connecting the centers of four of the coins (an example is shown). Now remove as few coins as possible until no square can be made.

Puzzle 13 Two Choice Question

生徒＼問題	1	2	3	4	5	6	7	8	9	10	得点
A	○	×	×	○	×	×	○	○	×	○	80
B	×	○	×	×	×	○	×	○	×	×	20
C	○	×	○	○	○	×	○	○	○	○	70
D	×	×	○	×	○	×	×	×	○	×	?

There are ten questions for which you can choose O or X as the answer. Each right answer is worth ten points. The results for Students A, B, and C are as shown in the table. But the teacher has forgotten to write down the total score for Student D, and he also has lost the correct answer for each question. From the table, find out D's score.

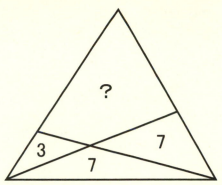

A triangle is divided into four parts by two straight lines as shown. The area ratio of the three parts is 3:7:7. What is the size of the fourth area?

Rock, Paper, Scissors

A boy and a girl played Rock, Paper, Scissors ten times. The boy used three rocks, six scissors, and one paper. The girl used two rocks, four scissors, and four papers.

There was never a tie, and the order in which the boy and girl used rocks, papers, and scissors is unknown. Who has won by how many wins?

(Puzzle by Yoshinao Katagiri)

Puzzle 16 Four Solutions of Four Pieces I

Cut this shape into four identical pieces. Don't stop after you find one solution—I've found four.

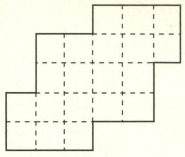

Puzzle 17 Four Solutions of Four Pieces II

Use the same rule as in Puzzle 16; try to find four solutions again.

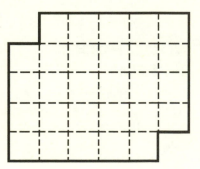

Puzzle 18
Square

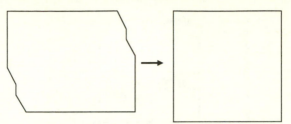

Cut the shape on the left into two parts and reconnect them to make a square.

Puzzle 19
Simultaneous Equations

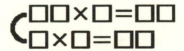

Fill in each of the squares with one each of nine digits 1 to 9 so that both equations are correct.

Even Number Sequence

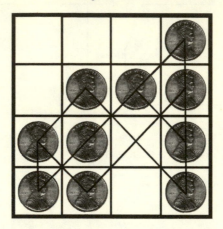

Ten coins are arranged on a 4 x 4 board, in the picture. As shown here, there are eight (horizontal, vertical, and diagonal) sequences where an even number of coins is lined up. Rearrange the ten coins so that you have the largest number of sequences with an even number of coins. Try to find an arrangement where you have the least number of sequences as well.

Age Guessing  Puzzle 21

Father: I have just realized that, if I switch the number in the ones digit and in the tens digit of my age, I get your age.
Son: Tomorrow you will be exactly twice as old as I.

How old are they as of today? Don't be too quick to answer.

Divide by Four  Puzzle 22

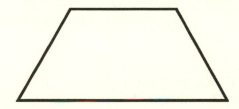

This shape is half of a regular hexagon. Divide it into four identical shapes; you can flip over shapes. I've found two solutions.

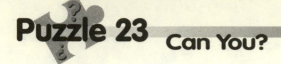

$$-101010$$

Can you move the "minus" to make this equal to nine fifty?

(Puzzle by Mel Stover)

Puzzle 24 To Have or Not to Have?

LAUGHING has it, but not CRYING.
HIJACK has it, but not TERRORISM.
FIRST has it, but not SECOND.
AFGHANISTAN has it, but not TAJIKISTAN.
CALMNESS has it, but not NOISE.
DEFINE has it, but not DECIDE.

What is it?

Five squares can be connected in 12 different ways, as shown in Figure 1. We consider two shapes to be the same if they become identical by flipping one of them over.

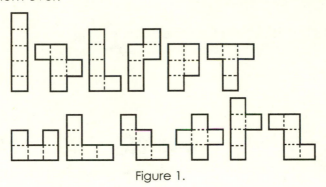

Figure 1.

Of these 12 shapes, find two different shapes that can fill the following shape in Figure 2. How many different solutions can you find? You are allowed to flip over the shapes.

Figure 2.

Puzzle 26 Pandigital (KOMACHI) Fraction I

$$\frac{\square}{\square\square} + \frac{\square}{\square\square} + \frac{\square}{\square\square} = 1$$

Fill out the squares in the equation using the numbers 1 to 9 once and only once. Two boxes together is a two-digit number.

Puzzle 27 Pandigital (KOMACHI) Fraction II

$$\frac{\square}{\square \times \square} + \frac{\square}{\square \times \square} + \frac{\square}{\square \times \square} = 1$$

Fill out the squares in the equation using the numbers 1 to 9 exactly once.

Triangulate a Square

Try to fill up this square using only right isosceles triangles (as shown). Each triangle should be a different size—try to use as few triangles as possible. So far, I know of two solutions.

Challenge
by Don Knuth

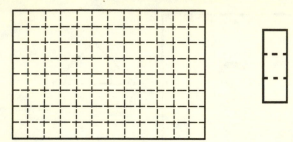

Fill an 8 x 12 grid using 32 1 x 3 pieces so that a "cross" is not formed. A "cross" is formed when two lines intersect at a four-way junction.

Note: Japanese tatami mats are arranged as in Examples I and II such that no line crosses the room. Tatami mats would never be arranged as in Example III. Don Knuth created this puzzle when he visited Japan and had a conversation about tatami mats with me. Since the given condition is severe, a solution can be found

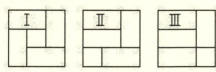

easily. If there are no restrictions, there are 51,493 different solutions. Thanks to this restriction, there are only two solutions.

Arrange the numbers 1–15 so that the sum of two neighboring numbers is always a square number.

Hint: You can find a solution using pencil and paper. The same square number will be used more than once.

(With help from Goro Tanaka)

Puzzle 31

Interlocking Pieces?

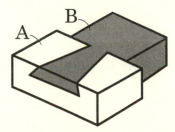

A white wooden piece and a black wooden piece are interlocked as shown.

Can you guess how they might come apart easily? There is no hollow space inside. The bottom view is just a black rectangle abutting a white rectangle.

Puzzle 32 Change of Heart

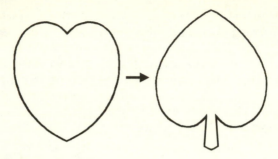

Divide this heart into three parts and rearrange them to make a spade.
(Puzzle by Sam Loyd)

Puzzle 33 Fill it Up with Coins

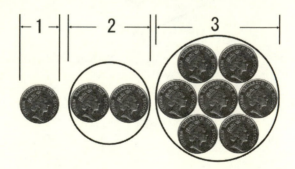

Two diameter 1 coins fit into a diameter 2 circle; seven diameter 1 coins fit into a diameter 3 circle. How many diameter 1 coins will fit into a diameter 4 circle?

LYM

Pick two numbers from 2–9. Create a number using the numerals for these two numbers in such a way that it can be divided by either of the two numbers. For example, 48 or 488 can be created using 4 and 8, and each of them can be divided by both 4 and 8. If you use 2 and 4, the smallest such number is 24. If you use 3 and 5, the smallest such number is 3555. Find two numbers so that the least such number is the largest of all the combinations.

This puzzle was named LYM (Least Yoshigahara Multiple) by Technology Review from M.I.T.

Bug House

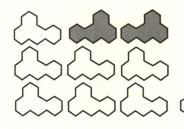
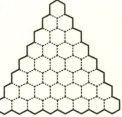

Place these nine bug-shaped pieces into the triangular honeycomb shape. You can rotate them, but you cannot flip them over. Notice that two pieces have already been flipped over: these cannot be flipped back.

Fill in the space by pencil. It is not so hard.

Puzzle 36 — What's Missing?

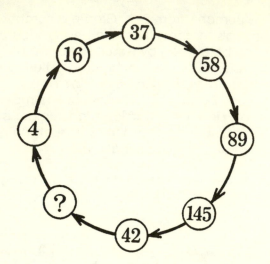

Eight numbers are placed in the circles as shown. What number is missing? Try to find the pattern.

Look at it carefully and you will begin to see the solution.

Connecting the Dots

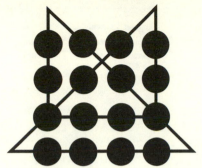

Black dots are arranged as a 4 x 4 grid as pictured. Connect the center of each dot in one stroke to form a loop. One solution is shown here; find another solution.

Equal Distance

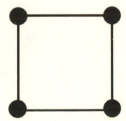

Consider the four vertices of this square. Each vertex is an equal distance from the two vertices to which it is connected.

Place nine points on a flat surface so that any point is an equal distance from three other points. Find one example.

Puzzle 39

Pallindrome Time

$$3 : 59 : 53$$

A palindrome is an expression which reads the same both forward and backward. Look at this clock. It shows exactly the same time by reading from left to right and from right to left(ignoring":"). On a 24 hour clock there are 660 times each day when we have such a time. Now find the following times:

1. The two palindrome times that are the closest.
2. The two palindrome times that are the farthest apart with no palindrome times between them.
3. The two palindrome times that are the farthest apart if other palindrome times are allowed between them.

Pandigital (KOMACHI) Time

My digital watch shows the time displayed here.

It reads August 19th, 23 hours, 46 minutes, 57 seconds. You will notice that it displays each of the numbers 1–9 once. There are 768 instances like this a year. Now, of such instances:

Find the Pandigital time that happens earliest in a year.
Find the Pandigital time that happens latest in a year.
When do we see a Pandigital time that contains the numerals 0–9?

TROMINO

Make a square or rectangle using many TROMINOES. Its smallest solution is 2x 3 with two pieces as shown above. But for this problem, such construction to make 2 x 3 is not permitted at any place. With this restriction, find the smallest rectangle or square.

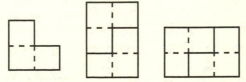

Illustrated by Ida Keiko

Puzzle 42 Switching Trains

跨線橋

We want to switch the Cargo Cars X and Y on this railroad. The Train L cannot go under the bridge on the sidetrack because its smokestack is too tall. The cargo cars can go under the bridge over the sidetrack. Find a way to exchange the cargo cars.

Puzzle 43 Building Highways

Four cities in a desert are placed 100 miles apart at the corners of a square.

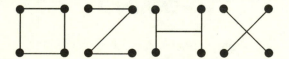

The four cities are to be connected by highways: Due to the small budget, the initial plan, Plan O, which was 400 miles in total length, needed to be shortened to Plan Z (341 miles), then to Plan H (300 miles), and finally to Plan X (283 miles).

Note: lengths are rounded to an integer number of miles.

One wise guy keeps saying there is even a shorter route for connecting four cities. What is it?

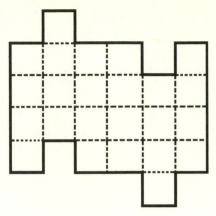

Figure 1.

Figure 1 shows a board with an unusual shape. It is easy to divide this into four identical parts (see Figure 2).

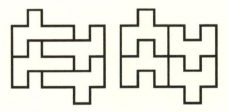

Figure 2.

Now, divide this board into three identical parts. I showed the 24 squares on the board to help you visualize the area.

Puzzle 45 All Gone?

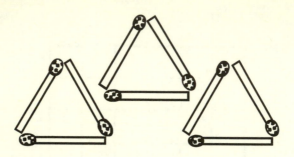

These three equilateral triangles are made of nine matchsticks. Move only two matchsticks and get rid of the triangles.

Puzzle 46 Change Direction

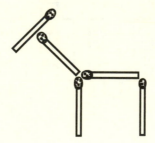

This horse is made of five matchsticks. Move only one matchstick and make the horse face the opposite direction.

Note: In the same way we speak about people, we would not say a horse faces the opposite direction if it simply turns its head 180 degrees (just to be clear).

(Puzzle by Mel Stover)

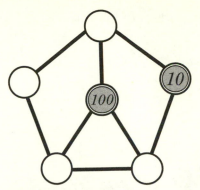

Suppose a 100 yen coin is used as a detective and a 10 yen coin is used as a fugitive. You, as the detective, go first by moving into an adjacent circle. The fugitive will move to an adjacent circle to escape from you. Find a way to catch the fugitive with the minimum number of moves.

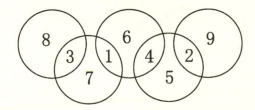

The numbers 1–9 are arranged as shown. If you add up the numbers in a circle, you get 11 for each of the circles. Rearrange the numbers so that the sum of the numbers in each circle is the same, but not 11.

Puzzle 49 Massive Multiplication

What is the product of the following 26 terms?

$$(x-a)(x-b)(x-c)\ldots(x-y)(x-z)$$

Puzzle 50 Cut and Rearrange a Cross

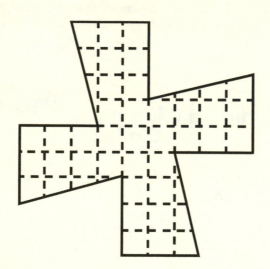

Cut this cross into four identical pieces and rearrange them to form a square.

The Curious Door Sign

PHUƧLUL9

Can you guess the meaning of this sign? It was on the glass door of a restaurant in Osaka.

Triangle Struggle

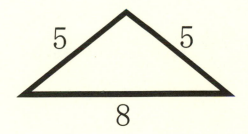

The length of the three sides of the triangle is 5:5:8. What is the area of this triangle?

Place the Coins

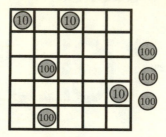

We want to put three 10 yen coins and five 100 yen coins onto a 5 x 5 board with one condition: Only the same kind of coins can be placed in a vertical, horizontal, or diagonal line. In the figure, we succeeded in placing five coins, but we cannot place the remaining three 100 yen coins. Rearrange the coins such that all eight coins are on the board.

Cut and Paste i

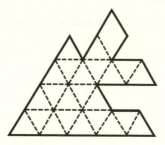

Using the dotted lines, cut this shape along some of the dotted lines into two pieces that you can arrange to make an equilateral triangle. You may flip the pieces over.

Cut and Paste ii

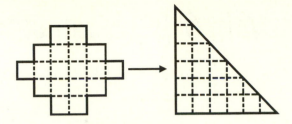

Divide the shape on the left into three parts. Rearrange them to make a right triangle like the one on the right. Don't be too square!

Trick Name Card I

My British friend James Jason works as a disk jockey at the BBC radio station, and his program is broadcast both on FM and AM at the same time.
His name card looks like this:

J.Jason DJ FM-AM

I've found a pattern hidden in this name card. James was shocked when he heard what I'd found. What was the pattern?

Puzzle 57 Trick Name Card II

This is my name card, which also has a hidden trick. What is it?

Puzzle 58 Uniform Division

1	2	3	4
5	6	7	8
9	10	11	12
13	14	15	16

The numbers 1–16 have been placed in a 4 x 4 grid, as shown. Cut the square at the thick line. The sum of numbers in each area is 68.

Find other ways to cut this square so that the sum of the numbers in the two areas is 68.

Three Contacting Matchsticks

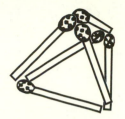

To arrange six matchsticks so that each end of each matchstick is in contact with exactly two others, you only need to create a regular tetrahedron. How would you satisfy this criterion if you arranged the sticks (you may use more than six) in a plane, i.e., in two dimensions?

Find the Rule

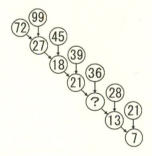

These numbers are arranged using a certain rule. Find the rule and fill in the blank circle.

Puzzle 61 **Typewriter**

These are ten alphabet letter keys as arranged on a typewriter. Find a ten letter word which you can type using only these keys.

Puzzle 62 **Seven Straight Lines**

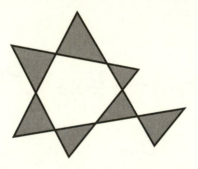

How many triangles can you make with seven straight lines? Try to make as many triangles as you can. The triangles are not allowed to overlap and cannot share edges. This figure shows only seven triangles.

Four and a Half Tatami Mats

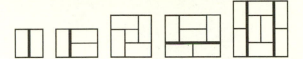

These figures show the standard way of arranging traditional Japanese tatami mats in a room.

Notice the room that has four rectangular mats and one square mat. No lines go across this room, while all the other rooms have a thick line going across them.

Using only the full tatami mats (rectangles), determine how many full tatami mats you need to fill either a square or a rectangular room such that there is no line going across the room. Try to use the least number of tatami mats possible.

Five Consecutive Numbers

① ☐☐ = ☐ + ☐ + ☐

② ☐ × ☐ = ☐ + ☐ + ☐

③ ☐ × ☐ = ☐ + ☐☐

④ ☐ + ☐ = ☐.☐ × ☐

⑤ ☐☐ = ☐☐ + ☐

Fill in the blank squares with five different digits. The five digits

need to be consecutive; for example, 1–5, 2–6, 4–8, or 5–9, etc.

Suppose we have a giant chocolate bar divisible into 4 x 6 pieces and we want to separate it into 24 pieces.

You may cut across the bar, but you may not cut two stacked pieces. In other words, you can only cut one layer of the bar along a line at a time. In order to make 24 separate pieces of chocolate, what is the minimum number of times you need to cut the bar.

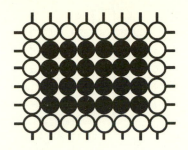

24 white Go stones and 24 Black go stones are arranged as shown. Form another rectangle of black stones which requires an equal number of white stones to surround it.

Coins in Two Dimensions

If we arrange four coins of equal diameter as a regular tetrahedron, each coin will touch three other coins. Now let's try this on a plane. What is the minimum number of coins you need so that each coin is touching three other coins on the plane?

Puzzle 68 Make a Square

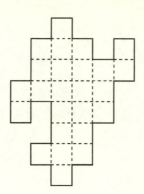

Separate this shape into three parts and rearrange them to form a square.

Puzzle 69 What Comes Next?

$$77 \Rightarrow 49 \Rightarrow 36 \Rightarrow 18 \Rightarrow ?$$

Equally Divided

Consider a plot of land in the shape of an equilateral triangle with 100 meters on a side. Divide this land into two parts of equal area using the shortest possible dividing line.

Wild Pig Maze

In this maze, go from the dark triangle to the white triangle. Pretend you are a wild pig and cannot turn unless you bump into a wall. Go ahead!

Puzzle 72 FlipIt

Four game pieces with black fronts and white backs are arranged in an area made of five squares. The goal is to turn all the black pieces white.

One piece can jump over pieces into an empty spot. The jumped pieces are flipped over, but the jumping piece is not.

A four-move solution is shown for an example problem.

Problem

| | ● | ● | ● | ● |

Solution

| | ←● | ● | ● | ● |

| ● | ○ | ○ | ○→ | |

| ● | ○→ | | ● | ○ |

| ←● | ● | ● | ○ | |

| ○ | ○ | ○ | ○ | |

Now solve two new problems. Try for the minimum number of moves.

① | ● | | ● | ● | ● | ② | ● | ● | | ● | ● |

A Shining Star  **Puzzle 73**

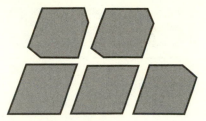

Rearrange these five pieces and make a star. (This is from a puzzle that appeared in a nineteenth century advertisement.)

 **Puzzle 74**

Escape from the Tower

An old queen, her daughter, and her son weigh 195, 105, and 90 pounds, respectively. The three have been captured in a room at the top of the tower.

The only way to escape from the tower is to use two baskets hanging over a pulley. When one basket is on the ground, the other basket is at the window of the tower. If the difference in weight is more than 15 pounds, the heavier basket goes down at such a speed that the person in that basket may not survive. You cannot control the speed of the basket.

Only one cannon ball is left in the tower room and it weighs 75 pounds. The cannon ball can survive the fall when the difference in weight between the two baskets exceeds 15 pounds. One basket can hold one person and one cannon ball.

How can you help the three of them escape from the tower in the least number of steps?

(Nineteenth Century Puzzle)

Puzzle 75 A Broken Calculator I

My broken calculator stopped showing the vertical lines. I typed an equation and got the answer above. Can you guess the equation? There are two possible answers.

Puzzle 76 A Broken Calculator II

I typed a number on the same broken calculator, and it showed three horizontal lines as pictured. Find a key operation that tells you what number I typed.

Switching Frogs

Three white frogs and three black frogs are put on squares as shown. The goal is to switch all the white frogs with the black frogs according to the following rules:

1. A frog can move into an abutting square if it is empty.
2. A frog can jump over one frog of the opposite color and move onto an empty square.
3. A frog cannot move diagonally, but it can move backward.

Puzzle 78 All Different

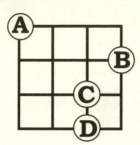

Four coins A–D are placed at lattice points of a 3 x 3 grid.

Using the Pythagorean theorem, the distance between coins is computed as follows:

A-B = $\sqrt{10}$
A-C = $\sqrt{8}$
A-D = $\sqrt{13}$
B-C = $\sqrt{2}$
B-D = $\sqrt{5}$
C-D = 1 .

As you can see, each distance is different.

Now, place five coins on Grid 1 and six coins on the Grid 2 at lattice points so that the distances between the coins are all different.

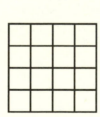

Grid 1.

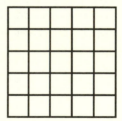

Grid 2.

Note: The Pythagorean theorem ($3^2 + 4^2 = 5^2$) makes it more difficult to solve this puzzle.

Odd Numbers and Even Numbers

"An odd number plus an odd number makes an even number. An even number plus an odd number makes an odd number. An even number plus an even number is an even number. Right?"

"Yes."

"An odd number times an odd number is an odd number. An even number times an odd number is an even number. Right?"

"Sure."

"An even number times an even number is an odd number. Right?"

"Huh?"

"You don't think so? An even number times an even number is an odd number."

"Why?"

Make a 10 x 10 grid with 2.5 cm x 2.5 cm squares and fill it with numbers as shown. Make the numbers small enough that you can hide them with a coin.

```
5 5 1 6 2 7 3 8 4 9
1 9 5 8 4 7 3 6 2 5
6 4 2 5 3 6 4 7 5 8
2 8 6 7 5 6 4 5 3 4
7 3 3 4 4 5 5 6 6 7
3 7 7 6 6 5 5 4 4 3
8 2 4 3 5 4 6 5 7 6
6 4 8 5 7 4 6 3 5 2
9 1 5 2 6 3 7 4 8 5
5 5 9 4 8 3 7 2 6 1
```

Suppose you say to someone, "Place one coin on any number you want while I am not looking." Then, you turn around and tell him what number he covered with the coin.

How did you do it?

Note: The answer that you have memorized all the numbers is not acceptable.

Puzzle 81

Time Calculation Using Pandigitals (KOMACHI)

Fill in the nine blank squares using each number 1–9 once. The minutes and the seconds should not be greater than 59.

□□分□□秒×□=□□分□□秒

[ ] [ ]minutes[ ] [ ]seconds × [ ] = [ ] [ ] minutes[ ] [ ] seconds

Puzzle 82

Time Calculation Using Pandigitals Plus 0 (OHMACHI)

The same rule applies here as in Puzzle 81, but this time use the digits 0–9.

① □□分□□秒×□=□時間□□分□□秒
② □分□□秒×□□=□時間□□分□□秒

1. [ ][ ]minutes [ ][ ]seconds × [ ] = [ ]hours [ ][ ]minutes [ ][ ]seconds
2. [ ]minutes [ ][ ]seconds × [ ][ ] = [ ]hours [ ][ ]minutes [ ][ ]seconds

How Many Loops?

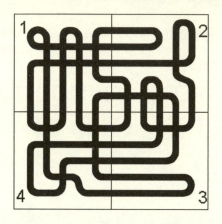

These four square pieces are connected so that there is one long loop. If we rearrange the squares, will the number of loops increase?

Cut out four big pieces from paper and try to rearrange them.

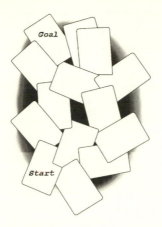

Rectangular pieces of paper are scattered. Start from the rectangle labeled "Start" and move to an adjoining rectangle. If you moved up onto the next piece of paper, then you must move down onto the next; i.e., you must alternate moving up and down while trying to reach the goal. You cannot move up or down twice in a row. Find a route with the smallest number of steps.

(With help from Mineyuki Uyematsu, Fujio Adachi, and Hiroshi Uchinaka)

Jumping Go Stones

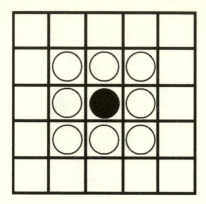

Eight white Go stones and one black Go stone are placed on the 5 x 5 Go board. Your goal is to leave one black stone in the center of the board. Use the following rules:

1. Each stone can jump over one stone horizontally, vertically, or diagonal.
2. A jumped stone is removed from the board.

This problem is not so hard.

Since every time one stone jumps over another the total number of stones will decrease by one, the number of jumps cannot be more than eight. Now, play the game with one more condition:

3. If one stone continues to jump, count it as one jump.

Now, try to find the solution with the smallest number of jumps.

This triangle has sides made up of three, four, and five matchsticks; its area is

$$(3 \times 4)/ 2 = 6.$$

Move five matchsticks and create a shape whose area is two.

Puzzle 87 One Place the Same

Five balls with different symbols are arranged in a pentagon as shown. Can you arrange the same five balls into a second pentagon so that, no matter how you move and/or flip it, you can never get more than one symbol to match with the first pentagon?

When we compare the two pentagons by flipping them over, moving them around, etc., they always have one place where the symbol of the balls matches.

Guess how the five balls are arranged in the second pentagon.

Note: The only acceptable configurations are where all five balls match; where only one of them matches; or where none of them match.

The Same Product — Puzzle 88

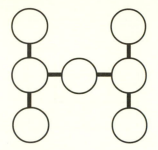

Choose seven digits from 1–9, each once only, and put them in the empty circles such that the products of the three numbers on each line are all the same.

No Squares — Puzzle 89

24 matchsticks are arranged as shown. The figure includes nine squares with sides of length 1; four squares with sides of length 2; and one square with sides of length 3, for a total of 14 squares.

Guess the smallest number of sticks you can remove so that there are no squares.

Puzzles 101

57

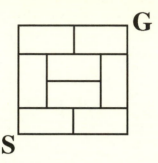

This town is a square with 400 meters on a side, and it has roads as illustrated. Each block is a 100 × 200 rectangle, and the blocks are arranged like Japanese tatami mats.

The person who lives at S wants to visit his friend at G. Since S would like to do a lot of exercise, he wants to take the longest route to visit G, but he cannot go on the same road more than once.

What is the longest distance he can take on a walk to see G?

$$[ ] \times [ ] = [ ][ ]$$

Put each of the numbers 1–4 in the blank squares. For example, you can use $3 \times 4 = 12$.

Where possible, fill in the following squares using the digits listed:

1–5	$[ ][ ] \times [ ] = [ ][ ]$
1–6	$[ ][ ] \times [ ] = [ ][ ][ ]$
1–7	$[ ][ ][ ] \times [ ] = [ ][ ][ ]$
1–8	$[ ][ ][ ] \times [ ] = [ ][ ][ ][ ]$
1–9	$[ ][ ][ ][ ] \times [ ] = [ ][ ][ ][ ]$

Find a ten-digit number that is made up of all different numerals and has the following characteristic: The number from the leftmost digit to the nth digit can be divided by n. For example, suppose a ten-digit number is ABCDEFGHIJ; the number ABC must be divisible by three; the number ABCDE must be divisible by five; and the number ABCDEFGHIJ must be divisible by ten. Find such a ten-digit number. It's not so hard.

The square is divided into five rectangles using four straight lines. The lengths of the horizontal and vertical sides of these 5 rectangles are all different and range from 1 cm to 10 cm. How was the square divided?

Hint: The square is 13 cm x 13 cm.

Puzzle 94 Height and Weight

A man says, "I am the average height and average weight of a Japanese man. Thus, I am an average man." However, he is still considered to be a little overweight. Why?

Suppose I want to find an integer whose rightmost digit becomes its leftmost digit when multiplied by four, while its other digits move one place to the right. Here is such a number:

102564 × 4 = 410256.

Now, find an integer whose rightmost digit becomes its leftmost digit when multiplied by six. Show the equation.

If you add up the nine numbers 1–9, you get 45. If you multiply the nine numbers
1–9, you get 362,880.

Now, suppose I have nine one-digit numbers. If I add all nine numbers and get 45, and if I multiply all nine numbers and get 362,880 as the total product, can I say that the nine numbers I have are 1, 2, 3, 4, 5, 6, 7, 8, 9?

Multi-Level Crossing

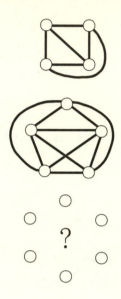

Four cities are connected by independent highways which do not cross, as shown. However, we cannot connect five cities without at least one multi-level crossing.

How many crossings do we need to connect at least six cities?

Squeeze-In! I

As shown in the figure, 2n circles of radius ½ fit in a 2 × n rectangle. Find the smallest n where 2n + 1 circles of the same radius fit.

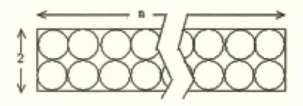

Squeeze-In! II

In a regular pentagon, try to squeeze in 23 isosceles triangles that have a top angle of 36 degrees and sides that are half the length of a side of the pentagon. Some space in the pentagon will not be filled.

Use construction paper to cut out the triangles and try to arrange them.

Puzzles 101

63

Puzzle 100 UN-Balance

Here is a scale with uneven arms.

Using two 5 pound weights, measure exactly 10 pounds of sugar.

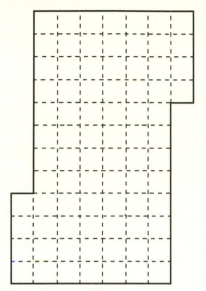

About twenty years ago, Professor Solomon Golomb proudly handed me this puzzle. He asked me to divide this shape into four identical shapes with the same area. I was not able to solve it at that time.

Professor Golomb assumed there was only one solution to this puzzle. Surprisingly, three solutions have been found in Japan since then.

This is the last puzzle for you. You will be considered to have completed this book if you find at least one solution.

SOLUTIONS

Solutions

There are six possible arrangements.

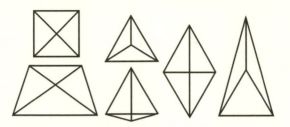

By the way, among four points, if all the lines connecting any two points are the same distance apart, then we cannot place the points on a flat surface; however, we can place them in three dimensions as the edges of a regular tetrahedron.

It takes the boatman at least 30 minutes. First, he takes Boats A and B (4 minutes) to the right bank, and returns to the left bank using A (2 minutes). He then takes Boats C and D (16 minutes) to the right bank and returns to the left bank using B (4 minutes). Finally, he takes Boats A and B (4 minutes) to the right bank.

Fold as shown.
Note: Two neighboring squares overlap.

Puzzle 4

None of the numbers contain an "e" when spelled out in English. For example: two, four, six, thirty, thirty-two, ... , sixty-six. Therefore, the next number will be 2000 (two thousand).

Note: I waited to use this puzzle until I sent my 1999 Christmas card and 2000 New Year's card—everybody who received one was surprised.

Solutions

Puzzle 5

This solution uses only two pieces.

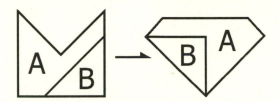

Puzzle 6

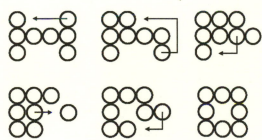

Puzzle 7

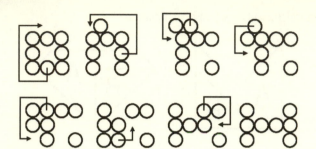

Puzzle 8

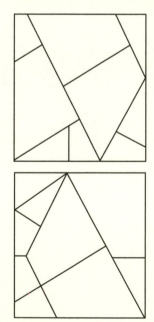

Puzzles 101

Here is the answer.

The mirror image of this solution is considered to be the same answer.

Do not try to make a triangle using numbers 1–21. No matter how hard you try, you cannot succeed.

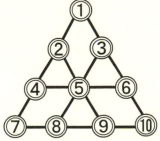

First remove Coin 2. Then take the following five steps: 7⇒2, 1⇒4, 9⇒7⇒2, 6⇒1⇒4⇒6, 10⇒3. At the third step, you can jump three times—resist that temptation, because you can solve the puzzle in fewer steps.

Puzzle 11

Solutions

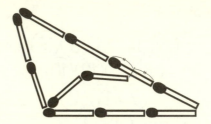

The horizontal matchstick inside the triangle is parallel to the base of the triangle, and its right end touches the center of one of the matchsticks on the side of length 4.

Here is another similar solution:

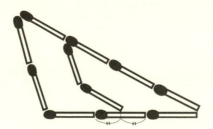

The minimum number of coins you can remove is six. Here is one example.

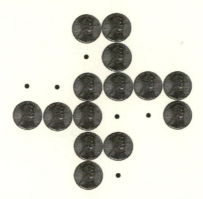

From the results of A, B, and C's answers, we can guess three possible patterns of correct answers—X, Y, and Z.

生徒＼問題	1	2	3	4	5	6	7	8	9	10	得点
X	◯	×	◯	◯	×	×	◯	×	×	◯	100
Y	◯	×	×	◯	◯	×	◯	×	×	◯	100
Z	◯	×	×	◯	×	×	◯	×	◯	◯	100
D	×	×	◯	×	◯	×	×	×	◯	×	40

We do not know which pattern is the correct answer. But the teacher will be relieved: D's score is 40 no matter which answer pattern he uses.

Puzzle 14

From

$$S_1:(S_2 + 7) = 3:7$$
$$S_2:(S_1 + 3) = 7:7,$$

we get

$$S_1 = 7.5 \quad \text{and} \quad S_2 = 10.5,$$

so

$$? = S_1 + S_2 = 18.$$

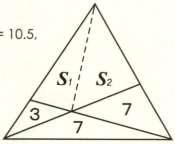

Puzzle 15

Since there is no tie with scissors, possible combinations are:

BOY	**GIRL**
Scissors 6	Rock 2
	Paper 4

Boy wins 4 and loses 2.
Then,

BOY	GIRL
Rock 3	Scissors 4
Paper 1	

Boy wins 3 and loses 1.

So, the boy wins 7 and loses 3 in total.

Puzzle 17

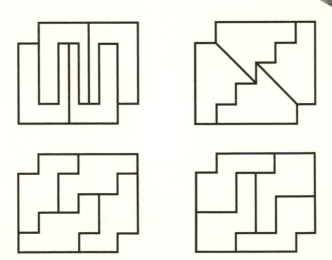

Puzzle 18

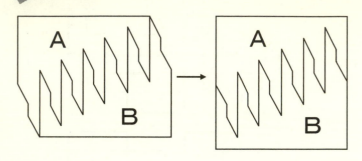

Puzzle 19

$$27 \times 3 = 81$$
$$6 \times 9 = 54$$

Solutions

The minimum is one; the maximum is 16.

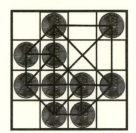

There are two solutions:

If tomorrow is the father's birthday: Father: 73 years old; Son: 37 years old.

If tomorrow is the son's birthday: Father: 52 years old; Son: 25 years old.

Puzzle 22

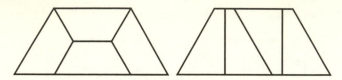

The solutions look simple, but they are difficult to discover.

Puzzle 23

The first thing that comes to mind is that nine fifty means 950. But the solution is 10 TO 10 (ten minutes to ten o'clock); in other words, nine o'clock, fifty minutes.

Puzzles 101

The words on the left all have three consecutive letters of the alphabet.

LAU**GHI**NG has it, but not CRYING
HIJACK has it, but not TERRORISM.
FI**RST** has it, but not SECOND.
AF**GH**ANISTAN has it, but not TAJIKISTAN.
CA**LMN**ESS has it, but not NOISE.
DEFINE has it, but not DECIDE.

I've found seven solutions:

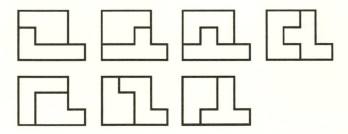

Puzzle 26

$$9/12 + 5/34 + 7/68 = 1$$

Note: Professor Donald Knuth of Stanford University is admired highly by computer programmers. He really liked this puzzle and became one of my fans. When he came to Japan to accept the Kyoto Prize, he even said that he wanted to meet me more than the Emperor! I was invited to his house in Stanford.

The Japanese TV program 3 Hours 45 Minutes broadcast this puzzle live. Only seven minutes passed before we received the solution to this puzzle by fax—I was impressed.

Puzzle 27

$$1/(3 \times 6) + 5/(8 \times 9) + 7/(2 \times 4) = 1$$

Puzzle 28

The solution on the right was found by Takao Yamaji.

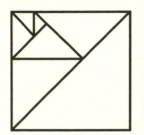 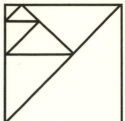

Solutions

Puzzle 29

Two similar solutions are:

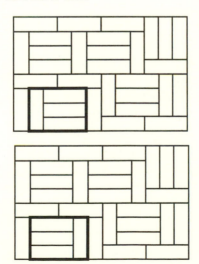

Puzzle 30

8-1-15-10-6-3-13-12-4-5-11-14-2-7-9

There is no solution for n smaller than 15.

Note: I tried to find an arrangement where the total of the first and last number is also a square number. I had to use a computer to find the solution.

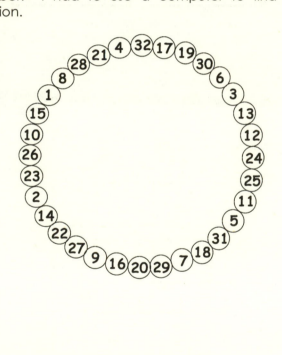

Puzzle 31

A and B look like this.

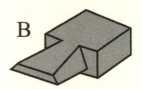

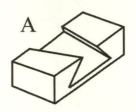

They slide together like this

Puzzle 32

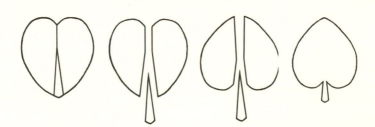

Puzzle 33

11 coins with some space left. The smallest diameter circle that can fit 11 coins is 3.924.

Puzzle 34

77777779779 is the smallest number which can be divided by both 7 and 9. 8888889888 is the next biggest answer.

Puzzle 35

This is the only solution.

Puzzle 36

You get the next number by squaring the digits of each preceding number:

$37 \Rightarrow 3^2 + 7^2 \Rightarrow 58 \Rightarrow 5^2 + 8^2 \Rightarrow 89$.
So, $42 \Rightarrow 4^2 + 2^2 \Rightarrow 20 \Rightarrow 2^2 + 0^2 \Rightarrow 4$,

so 20 is the answer.

Puzzle 37

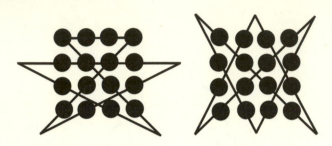

Puzzle 38

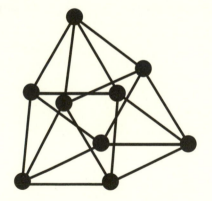

If this structure is made of pins connected at their ends, it is flexible, so there are an infinite number of solutions. For an animation of this structure, see http://theory.lcs.mit.edu/~mdemaine/nob_linkage/.

Solutions

1. 9:59:59 and 10:00:01. (only 2 seconds between them)

2. 15:55:51 and 20:00:02. (4 hours, 4 minutes, 11 seconds between them)

3. 1:33:31 and 13:33:31. (The farthest apart two times can be is 12 hours. For example, between 0:00:00 and 23:55:32, only 4 minutes, 28 seconds elapse.)

Puzzle 40

1. 3:26   17:48:59   March 26th, 17 hours, 48 minutes, 59 seconds.

2. 9:28   17:56:43   September 28th, 17 hours, 56 minutes, 43 seconds.

3. When the watch is broken. For example: 12:39 07: 58:46.

$12:39:07:58:46$

Puzzle 41

Eight
Pieces

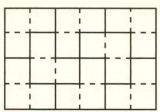

Puzzle 42

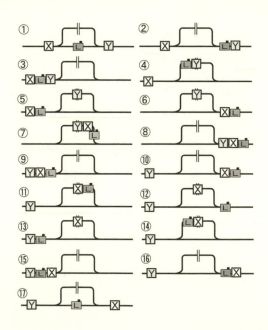

Puzzles 101

This plan is called the Hyper X Plan since it is between H and X. The total length is 281. All angles are 120°.

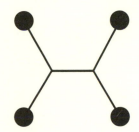

This is not a sneaky solution!

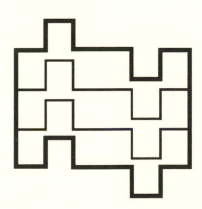

Puzzle 45

This will get rid of the triangles!

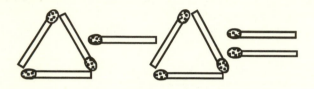

Puzzle 46

If you move the matchstick as in Step 2, the horse is now facing right, as in Step 3.

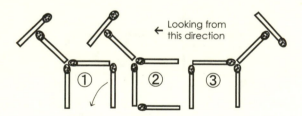

Looking from this direction

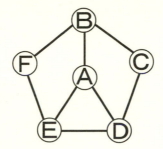

A⇒E, C⇒B, E⇒D, B⇒F, D⇒A. After that, you can catch the fugitive whether he moves to B or E. The minimum number of moves is six.

Here are three solutions. The sum of each circle is 14, 13, 13, respectively.

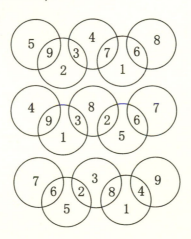

 Puzzle 49

The answer is 0 because the product includes (x- x) near the end.

A professor wrote down x^{26} and thought it over. He couldn't solve it.

Puzzle 50

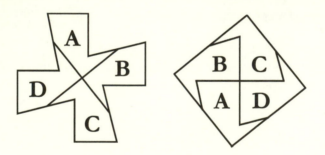

Note: This cross resembles a Maltese cross.

PULL
HꙄUꟼ

There was a sign PULL facing us and there was also a sign PUSH on the other side of the glass door facing the other way. Because the letters were spaced apart, this is a confusing sign.

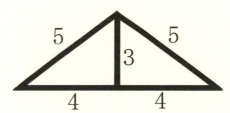

The area is 12. By drawing the altitude at the center, you can see that there are two triangles whose sides are of length 3:4:5—a Pythagorean triangle. So the answer is $(4 \times 3)/2 \times 2 = 12$.

Solutions

Puzzle 53

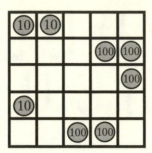

This is the only solution to this puzzle—reversed and rotated versions of this answer are not considered different.

Puzzle 54

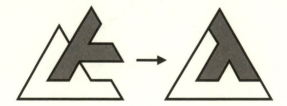

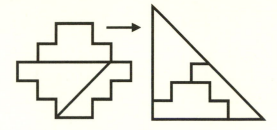

The first letter of each month appears in order: June, July, August, September, October, November, December, January, February, March, April, May.

It may be difficult to see, but you can read it as NOBUYUKIYOSHIGAHARA. It also reads the same turned upside down.

Super talented prodigy Scott Kim of California created it for me. It seems that some names are more difficult to invert than others. He came up with this kind of tricky writing for my wife and my daughter in a mere three minutes, but it took him 20 years to come up with one for my name.

Puzzle 58

Solutions

1	2	3	4
5	6	7	8
9	10	11	12
13	14	15	16

1	2	3	4
5	6	7	8
9	10	11	12
13	14	15	16

1	2	3	4
5	6	7	8
9	10	11	12
13	14	15	16

1	2	3	4
5	6	7	8
9	10	11	12
13	14	15	16

1	2	3	4
5	6	7	8
9	10	11	12
13	14	15	16

1	2	3	4
5	6	7	8
9	10	11	12
13	14	15	16

This is one of the solutions.

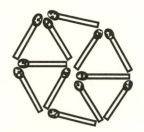

Note: I changed this puzzle to "four contacting matchsticks" and submitted it as a quiz for a prize to the Japanese Scientific magazine called *QUARK*, published by Kodansha Co. The smallest number of matchsticks used to make such an arrangement was 104; the answer was submitted by Professor Hiroshi Yabe, who won first prize. It required a complicated computation, so I will just show the solution.

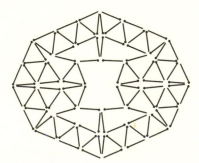

The second place winner used 108 matchsticks, and the third place winner used 114 sticks. I had a solution using 112 matchsticks, but the computation was incorrect, so I was disqualified.

Puzzle 60

The answer is 12. You get 27 by adding 9 + 9 + 7 + 2, and 18 by adding 4 + 5 + 2 + 7.

According to this rule, you get 12 by adding 3+6+2+1.

Note: If the number at the end was 15, then the solution you first thought of would have been correct solution. If you were not careful, you would not have noticed that this answer is the wrong solution.

As a maker of tricky puzzles, I consider this my masterpiece.

Puzzle 61

TYPEWRITER

Note: I did not say to find a word which you can type using each of these keys.

Puzzle 62

Here are eleven triangles!

Solutions

98

Arrange 15 mats as shown.

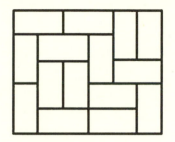

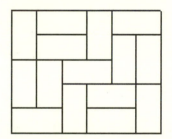

1. $12 = 3 + 4 + 5$
2. $2 \times 6 = 3 + 4 + 5$
3. $8 \times 9 = 7 + 65$
4. $3 + 6 = 4.5 \times 2$
5. $3\wedge4 = 75 + 6$

Puzzle 65

23 times.

No matter how you cut a sheet, the number of sides increases by one as you cut. To have 24 pieces, you need to cut 23 times. No more, no less.

Puzzle 66

30 stones, arranged as shown.

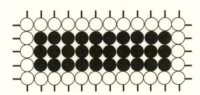

Note: I once made a three-dimensional version of this puzzle using sugar cubes, which filled up an 8 x 10 x 12 box. I approached a confectionary company to market this box saying "This box looks like it is made of all white sugar cubes, but inside there is the same number of red cubes. There Are 480 pairs of the white and the red sugar cubes. It may be a perfect present to give away at a wedding (since red and white are considered lucky colors)." But I did not receive any offer from this company.

In the three-dimensional version of this puzzle, there are 20 different solutions.

Solutions

Puzzle 67

The minimum number is 16.

Puzzle 68

Thanks to Pythagoras, this is not cheating.

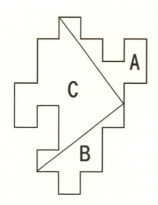

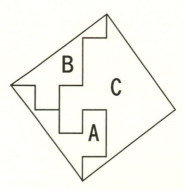

Solutions

Puzzle 69

If we interpret 77 as 7 times 7, you get 49. Then, 4 times 9 is 36; 3 times 6 is 18; and 1 times 8 is 8. So the answer is 8.

Note: In England, I bought a long tube-shaped candy which contained this puzzle inside. Unfortunately, it started melting, so I ate it.

Puzzle 70

Using one-sixth of a circumference is the shortest.

The direct line is not necessarily the shortest.

Note: We will not worry that such a division will make it harder to use the land.

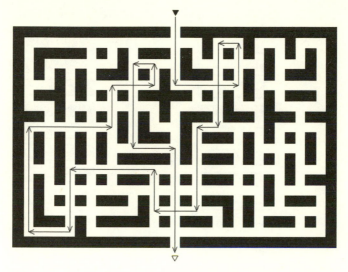

Note: Use this as a New Year's greeting card for the year of Boar.

Puzzle 72

1. Nine moves.   2. Ten moves.

① 9手   ② 10手

Note: I expanded this puzzle into a product that was marketed in the U.S., called FlipIt, which became tremendously popular.

104

How about this? It is not cheating.

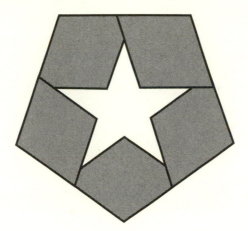

Puzzle 74

1. The cannon ball goes down.
2. The son goes down; the ball goes up.
3. The daughter goes down; the son goes up.
4. The cannon ball goes down.
5. The queen goes down; the daughter and the cannon ball go up.
6. The cannon ball goes down.
7. The son goes down; the cannon ball goes up.
8. The cannon ball goes down.
9. The daughter goes down; the son goes up.
10. The son goes down; the cannon ball goes up.
11. The cannonball goes down.

Puzzle 75

The first one is easy:

$43 \times 9 = 387$,

but the next one may take longer to find:

$143 \times 9 = 1287$.

Puzzle 76

Here is one way to find out:

Divide the number by two. From the answer, you can distinguish six numbers.

Display Number

表示　数

.	→ 2
.≡	→ 3
≡.≡	→ 5
≡.	→ 6
-.	→ 8
-.≡	→ 9

There may be other solutions.

1	2	
3	4	5
	6	7

Here is one of the many solutions; Fifteen is the smallest possible number of moves.

$5 \Rightarrow 4, 3 \Rightarrow 5, 4 \Rightarrow 3, 6 \Rightarrow 4, 7 \Rightarrow 6, 5 \Rightarrow 7, 4 \Rightarrow 5, 2 \Rightarrow 4,$
$1 \Rightarrow 2, 3 \Rightarrow 1, 5 \Rightarrow 3, 4 \Rightarrow 5, 6 \Rightarrow 4, 2 \Rightarrow 6, 4 \Rightarrow 2.$

Puzzle 78

Here is the one of the solutions for each grid.

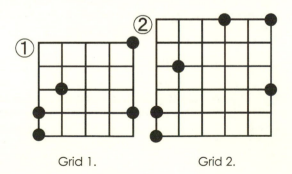

Grid 1. Grid 2.

Solutions

"An odd number plus an odd number" has 26 letters—an even number.

"An even number plus an odd number" has 27 letters—an odd number.

"An even number plus an even number" has 28 letters—an even number.

"An odd number times an odd number" has 27 letters—an odd number.

"An even number times an odd number" has 28 letters—an even number.

But, "An even number times an even number" has 29 letters—its odd, isn't it?

Note: This puzzle works in Japanese as well.

Puzzle 80

Pick any square consisting of four numbers which includes the hidden number. For example, in the figure, suppose the 5 in the center is hidden. Pick any of the squares consisting of 4385, 3557, 8552, or 5726.

4	3	5
8	5	7
5	2	6

We find that $4 + 3 + 8 + 5 = 3 + 5 + 7 = 8 + 5 + 5 + 2 = 5 + 7 + 2 + 6 = 20$. In other words, to find the hidden number, add the three other numbers in a square containing the hidden number and subtract the total from 20.

18 minutes 49 seconds × 3 = 56 minutes 27 seconds.

Note: I once published the puzzle "Fill in the Squares" on an advertising poster in the Osaka Subway" (repetition is allowed):

[ ] [ ] [ ] [ ] [ ] [ ] × [ ] [ ] [ ] [ ] [ ] [ ] = 123,456,789.

The puzzle was too difficult and there were lots of phone calls doubting whether there is actually a solution. To keep your peace of mind, here is the solution:

10,821 x 11,409 = 123,456,789

Puzzle 82

1. 50 minutes 42 seconds x 9 = 7 hours 36 minutes 18 seconds
2. 5 minutes 09 seconds × 84 = 7 hours 12 minutes 36 seconds

Note: Here is an amazing 0–9 computation. I am not sure how anybody could have found this:

286,794 × 5,103 = 1,463,509,782 = 479,682 × 3,051.

Solutions

Puzzle 83

As you can see, you can create any number of loops between 1 and 6 by rearranging the squares.

1 loop

2 loops

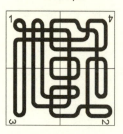

3 loops

4 loops

5 loops

6 loops

Solutions

110

Puzzles 101

Puzzle 84

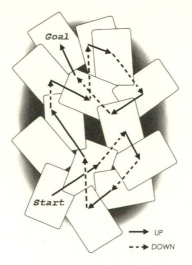

→ UP

- - → DOWN

Although there are only 14 pieces of paper, it took 15 steps.

Note: I found the original idea of this puzzle in England, the country of mazes, in 1999, and restructured it into this puzzle.

Puzzle 85

Solutions

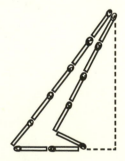

1) $1 \Rightarrow 2 \Rightarrow 3 \Rightarrow 4 \Rightarrow> 5$
2) $6 \Rightarrow 7$
3) $8 \Rightarrow 9 \Rightarrow 10$
4) $5 \Rightarrow 1$

Four moves is the minimum number.

Puzzle 86

How about this?

112

Puzzles 101

Here is the order.

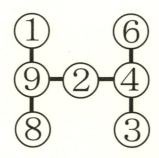

It is easy to guess that 5 and 7 cannot be used:

$$1 \times 9 \times 8 = 9 \times 2 \times 4 = 6 \times 4 \times 3 = 72.$$

Puzzle 89

Here is one of the solutions. More than five matchsticks need to be removed.

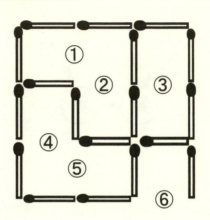

Puzzle 90

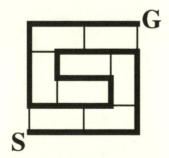

2,400 meters. It looks easy to find, but it actually takes a long time to figure this out.

$$13 \times 4 = 52$$
$$54 \times 3 = 162$$

There is no solution for digits 1–7.

$$582 \times 3 = 1746$$
$$1738 \times 4 = 6952$$

Note: There may be more solutions. If an equation is correct, it is a solution.

Puzzle 92

3816547290

This is one of the solutions.

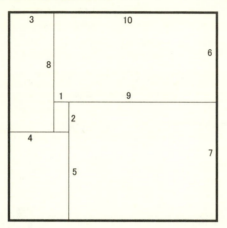

Consider a wooden cube which is 1 cm on a side and weighs 1 gram. A wooden cube which is 2 cm on a side weighs 8 grams. The average length of these two wooden pieces is 1.5 cm and the average weight is $(1+8)/2 = 4.5$ grams.

On the other hand, the weight of a wooden cube with 1.5 cm on a side would be 3.375 grams. So, if it weighs 4.5 grams, which is the average weight, it is overweight.

Solutions

The smallest possible number has 58 digits.

**1016949152542372881355932203389830508474576271186
440677966 × 6 =
610694915254237288135593220338983050847457627118
644067796**

The minimum number of digits for such a number multiplied by n is: for each "times n" is n = 2 $\Rightarrow$-> 18 digits; n = 3 $\Rightarrow$ 28 digits; n = 4 $\Rightarrow$ 6 digits; n = 5 $\Rightarrow$ 6 digits; n = 6 $\Rightarrow$ 58 digits; n = 7 $\Rightarrow$ 22 digits; n = 8 $\Rightarrow$ 13 digits; n = 9 $\Rightarrow$ 44 digits.

Note: Since the algorithm is simple, it may be faster to get the solution if we compute by hand. A middle school student spent one solving this puzzle during class; afterward, he got a reward, and then he paid even less attention in class!

Not necessarily. There is another combination of numbers. Using factorization into primes, we get

$$362880 = 2^7 \times 3^4 \times 5 \times 7.$$

You need to find nine one-digit numbers whose sum is 45. Since you need to have 5 and 7 in the product, you only need to find the 7 numbers whose total is 33:

$$45 - (5+7) = 33.$$

You will discover that 1, 2, 4, 4, 4, 5, 7, 9 also have the same characteristics.

Puzzle 97

Three crossings are needed.

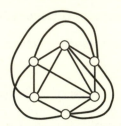

Note: For your information:

# of Cities	1	2	3	4	5	6	7	8	9	10
# of Crossings	0	0	0	0	1	3	9	18	36	60

Solutions

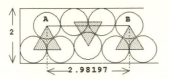 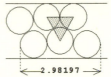

Drawing 1   Drawing 2

Solutions

As shown in Drawing 1, we arrange sets of three circles in an "equilateral triangle configuration." We then pack them tightly, starting against the left edge and alternating the orientation of the triangles. We computed the distance between A and B to be approximately 2.98197 using the Pythagorean theorem. The arrangement shown in Drawing 3 using three circles (1, 2, 133) and 55 copies of the six-circle arrangement in Drawing 2, puts 55 x 6 + 3 = 333 circles into a 2 X 165.9993 rectangle.

Conclusion: When n = 166, a 2 x 166 rectangle can contain 2n + 1 (333) unit circles.

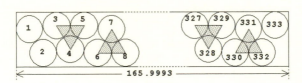

Drawing 3

Puzzle 99

There are three solutions. Here is one of them.

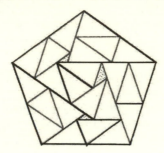

Puzzle 100

Put 10 pounds of weight (5 lbs + 5 lbs) on a plate, and put sugar on another plate until it balances. Then, remove the weights and place sugar there until it balances again: The sugar we have now is exactly 10 pounds.

When I was 10 years old, I created this problem and sent it to a newspaper for boys and girls, and won a Grand Prize.

Solutions

The leftmost solution is the solution Professor Golomb had in mind.

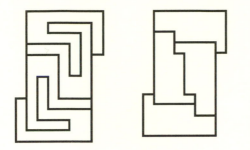
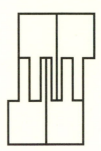

Note: The discovery of these solutions has made Japan, formerly unknown in the puzzle community, famous. People now know out about Japan's talent for solving puzzles. Puzzles connect people all around the world—I call the network among the puzzle community the Puzzletopia. If you find a fourth solution to this puzzle, please let us know. You will be famous.